Exploring the Potential Improvement of Quality Control in the Construction Industry with the Use of Digital Technology

Chris Staples

ELIVA PRESS

ELIVA PRESS

Chris Staples

This book is aimed at critiquing the application of digital technologies for the potential improvement of quality management in the construction phase of civil engineering projects.

To address the aim, two objectives emerged. The first was reviewing digital technologies that were available for use in quality control that could assist in the reduction of defects. The second was seeking the viewpoint from construction professionals to develop a quality management framework employing the most applicable digital technologies.

In support of the objectives, a qualitative research approach involved multiple sources of data collection, gained from literature and interviews. Participants in the interviews included digital engineering specialists, designers, main contractors and subcontractors. The data accumulated was analysed from both the answers to direct questions and by coding to provide a quality management framework of the most applicable technologies.

Published: Eliva Press SRL
Address: MD-2060, bd.Cuza-Voda, 1/4, of. 21 Chişinău, Republica Moldova
Email: info@elivapress.com
Website: www.elivapress.com

ISBN: 978-1-952751-39-4

Abstract

The construction industry has traditionally relied on paper to manage quality records, which has resulted in data analysis being challenging. To overcome this issue, the industry has started deploying digital-collaboration and field-mobility solutions.

This research was aimed at critiquing the application of digital technologies for the potential improvement of quality management in the construction phase of civil engineering projects.

To address the aim, two objectives emerged. The first was reviewing digital technologies that were available for use in quality control that could assist in the reduction of defects. The second was seeking the viewpoint from construction professionals to develop a quality management framework employing the most applicable digital technologies.

In support of the objectives, a qualitative research approach involved multiple sources of data collection, gained from literature and interviews. Participants in the interviews included digital engineering specialists, designers, main contractors and subcontractors. The data accumulated was analysed from both the answers to direct questions and by coding to provide a quality management framework of the most applicable technologies.

The key findings showed that implementation of digital engineering for quality assurance was at an early stage of development. Applicable digital quality applications were identified as electronic document management systems, personal digital assistants, building information modelling (BIM), mobile construction application products (apps), clash mitigation using BIM, real time performance information, point clouds of as-built construction, three-dimensional vision on mobile phones and barcodes.

Implications for practice indicated that training, visible use of digital technologies on site and the provision of an effective common data environment were paramount in instigating digital applications. Also, academia could assist in providing this common data environment. Once the appropriate technologies, readily available from vendors, were appreciated by company management, the most significant outcome was company willingness at director level for the adoption of digital engineering. The originality and value of this research derived from there being limited studies considering a unified approach, utilising the available digital processes, for quality assurance in the construction phase of civil engineering works.

Contents

Chapter 1. Introduction

1.1 Context

In recent decades, the construction industry systems for improving and managing quality have evolved through four discrete stages: inspection, quality control, quality assurance to total quality management, whereby now a process of continuous and company-wide improvements is employed (Dale, 2006). Architects, consulting engineers, main contractors, specialist sub-contractors and suppliers come together on the same project, but each will be simultaneously involved with other projects (Cheetham and Carter, 1993).

The success of a project can be evaluated by the degree to which it meets the customer's requirements. Achieving this aim requires not only the resources of many organisations and individuals, but also the successful interaction amongst these parties. The concept of quality assurance has arisen to ensure that customer requirements within a defined level of quality and conformance are achieved (Chan, 2011). Quality assurance is provided through the implementation of systematic management techniques ensuring control of the activities carried out by each party.

One of the best-known quality assurance certification systems is the ISO 9000 series that was originally published in 1987. It focuses on companies documenting their quality systems in a series of manuals, to facilitate good practice through supplier conformance. ISO standards provide a baseline against which an organisation's quality can be judged via multi-disciplinary participation in quality improvement efforts, documentation of procedures and the basic structural elements necessary for assurance systems (Maguad, 2006).

The construction industry still relies mainly on paper to manage its processes, such as defects lists and quality records. Due to the lack of digitisation, information sharing is delayed and may not be universal. Clients and contractors therefore often work in different versions of reality. The use of paper makes it difficult to capture and analyse data, preventing historical performance analytics that leads to better outcomes (McKinsey, 2016). As well, paper trails take longer to execute. To overcome this issue, the industry is beginning to deploy digital-collaboration and field-mobility solutions.

With the advent of digital technology, using mobile or tablet-based devices on site, engineers can access digital models of infrastructure projects and quality assurance data (Laing O'Rourke, 2018). The technology enables visibility across a project, permitting companies to view and interrogate the status of each component from its initial design to manufacture, installation and onto final acceptance.

Digital engineering in the form of building information modelling (BIM) has transformed design, though digital engineering is yet to be fully applied in processes concerning site works within the ISO 9000 family of standards. It is noted that the aspiration of BIM represents full collaboration between all disciplines by means of a single, shared project model which is held in a centralised repository (NBS, 2018).

For the potential improvement of quality assurance in the construction industry with the use of digital technology, there are two main elements to this dissertation. The first is the review of digital technologies that are available to the construction industry for use in quality control during the construction phase. The second is seeking advice from construction professionals to develop a quality management framework, employing the most applicable digital technologies from the initial appraisal. The intention is this digitally connected approach to quality assurance can be delivered by harnessing the specialist technologies and capabilities of business intelligence, BIM, surveying, geographic information systems, automation, collaboration and digital mobility.

1.2 Research Aim

The research aim is undertaking a comprehensive explorative analysis of the potential improvement of quality control in the construction industry with the use of digital technology.

The research addresses the provision of proactive construction quality management using appropriate digital engineering technology, as currently there is often insufficient quality control information at the very point when a component of a building is being created. This component creation may involve both design issues involving the buildability of co-ordinated drawings and the physical construction of the building itself.

1.3 Research Objectives

The research objectives are listed as follows:

- Conduct a review of current commercial technologies available in the construction sector that could be greater utilised in quality management
- An assessment process to determine what technologies have the greatest potential to improve quality management
- The changes required in processes within quality management systems to accommodate this technology
- The creation of a framework of technologies for digital quality management in civil engineering. This framework takes advantage of digital technologies that to date are typically unexploited in construction quality processes

1.4 Research Questions

The research questions focus on the delivery of a digitally connected approach to the management of construction quality assurance.

The following are the main questions:

- What are the principal contract documents that bind the quality assurance processes together?
- Does the application of BIM assist digitally connected quality assurance?
- How do we ensure staff and subcontractors reap the potential benefit of digital engineering?
- Which suite of the digital technologies indicated by interview participants is the most appropriate?

The primary research involves a literature review and interviews. Secondary research is developed from reports and studies by government agencies, trade associations and digital engineering companies.

Due to the structure of the construction industry containing designers, main contractors and subcontractors (each within contractual arrangements), the research also addresses the approach of personnel from these three principal groups. As well within these groups, the research considers the motivation of people and their organisation in relation to digital engineering.

1.5 Criteria of Study

The research output is focused on the infrastructure sector of civil engineering. This sector of the industry has the resources and technical acumen to apply digital technology in its initial use on site. The search for suitable digital quality assurance systems utilises a worldwide literature review. In principal, the criteria for this research considers the application of digital quality processes over the next 4 years during the construction phase for civil engineering projects of over £50 million expenditure.

1.6 Methodology

For this dissertation, the qualitative research uses a relativist approach, whereby the views of participants are utilised to discover the appropriate digital applications for construction phase quality assurance. The qualitative research involves multiple sources of data collection gained from literature and interviews. This is followed by analysing the data to provide a quality management framework of the most applicable technologies. The research gathers information by asking participants to answer open ended questions (Flick, 2009). Participants are chosen from designers, main contractors, subcontractors and specialists in digital engineering. The criteria of selection are further based on knowledge and experience of construction management and quality assurance. The same questions are asked to all interviewees to obtain quantifiable data on further investigation. The meetings are transcribed into text to ensure accurate records are available for analysis of the data. From the responses received the research seeks broad patterns and themes.

For elements of the research qualitative data analysis coding is carried out to seek structure to recorded observations. This involves developing specific codes for types of participant responses and quantitively organising the resultant data.

This approach provides a means to introduce interpretations into the qualitative research methodology.

A spreadsheet, comprising a series of tables, is then employed to organise and present insights from data received, utilising the coded groups of topics from text transcripts. The output from the direct questions in the interviews to discover the most appropriate digital quality technologies is placed in the same spreadsheet. The most popular processes chosen by the interviewees, presented in Chapter 4 Results and Discussion, being the digital technologies selected for quality assurance in civil engineering projects. In the following Chapter 5, the implications of this research to civil engineering infrastructure projects for both practice and academia are outlined.

Ethical research practices are considered and reviewed whilst instigating the dissertation. This involves respecting sources, preserving data that runs against the results of findings, asserting claims only as strongly as warranted and recognizing the limits of certainty (Booth et al, 2008).

1.7 Dissertation Structure

Chapter 1 **Introduction:** (this Chapter) The introduction has provided a high-level summary of the context for the application of digital engineering in quality assurance, followed by the research aim, objectives, research questions, criteria of the study and methodology

Chapter 2 **Literature Review:** The Chapter reviews published literature commencing with quality control in the construction industry, leading onto the adoption of digital technology, followed by digital engineering, the digital engineering environment and finally digital engineering quality applications

Chapter 3 **Methods and Data Collection:** This section describes the philosophical approach, the methodology, the research design and methods to address the research questions

Chapter 4 **Results and Discussion:** This Chapter presents and discusses the results of the research, the findings, the outcome analysis, plus the synthesis with literature reviewed

Chapter 5 **Implication:** This part outlines the implications of this research to civil engineering infrastructure projects for both practice and academia

Chapter 6 **Conclusions:** The Chapter covers the summary of conclusions from this research, its limitations, then contributions to the body of knowledge and finally recommendations for future investigation are provided

Chapter 2. Literature Review

This Chapter initially reviews quality assurance, then evaluates digital engineering literature and concludes with a table of possible digital quality management processes. This literature appraisal is approached by considering firstly quality control in the construction industry and secondly the adoption of digital technology with the evaluation of digital engineering, its working environment and digital engineering quality applications.

2.1 Quality Control in the Construction Industry

A common definition of quality is conformance to requirements or specifications, with a more general description being quality is fitness for use (Wadsworth et al, 2002). Quality management is now seen as a process that can be universally applied throughout a business and that by implication is the responsibility of all managers in the operation. It applies to all parts of the organisation (Slack et al, 2006). As suitably stated in this article, the impairment of quality is shared by all, so everyone also can provide improvements.

For example, in the automotive industry the Japanese company Toyota, ensures effective quality combined with low cost, short lead times, good safety and high morale through the elimination of waste. For the multinational engineering conglomerate, General Electric defect reduction in the form of Six Sigma is a data-driven approach, centred around six standard deviations between the mean and the nearest specification limit in any process (Liker, 2004). Though an awareness of these renown approaches is appreciated in the construction industry their application is limited.

The 'Made Smarter Review', initiated by the United Kingdom Industrial Strategy Green Paper (Made Smarter, 2017), promotes the utilisation of industrial digital technologies. This approach should be positive for employees and can lead to improved job satisfaction through the replacement of dull and repetitive tasks. Capital returns and quality should improve through increased accuracy and repeatability. Additionally, in a Presidential Address from the Institution of Civil Engineers entitled Engineering a Digital Future (Broyd, 2017), it is mentioned that as borne out in practice digital engineering can deliver projects at reduced cost, on time, with a quality and precision that provides the operation and management of truly smart infrastructure.

Reductions in construction, operational and maintenance costs can be provided by building information modelling. Also, improved performance and quality can be achieved by effectively using this modelling data (Davila Delgado et al, 2017). However, considered in isolation, the raw data has little use and value. This data must be processed and put into a geometric context within the infrastructure asset, facilitating the interpretation and analysis of the information. It is noted that full application of BIM beyond the design stage and into the construction phase is still in its infancy, with Bryde et al (2013) exploring the extent to which the use of BIM has resulted in reported benefits on a cross-section of construction projects.

2.2 The Adoption of Digital Technology
2.2.1 Digital Engineering
With the advent of digital engineering, utilising devices such as tablets and smart phones that take the technology out into the field, industry is evolving rapidly and irreversibly. The traditional command and control project management approach is becoming obsolete. Within this developing framework, digital engineering represents a great challenge in terms of controlling design and decision-making. Information now arrives continuously and no longer is owned by a few people (Laing O'Rourke, 2018). This new operating structure cuts across cultures, necessitating a standardisation of skills and qualifications. Upon review it is seen as a global phenomenon, particularly dependent on emerging software provided by vendors.

BIM, a collaborative manner of working, allows for more efficient methods of designing, delivering and maintaining physical assets throughout their entire lifecycle (BSI, 2017). As well 4D planning, the process of combining a 3D BIM model with a project's construction programme, enables pre-construction and continual review of the building process through the production of visual simulations. The concept of BIM has risen rapidly in the field of construction engineering management (Li and Yang, 2017). The aspiration of Level 3 is the term when BIM represents the full collaboration between all disciplines using a single project model.

The ability to process large amounts of data with the extraction of useful insights has revolutionised society. This phenomenon is known as 'big data'. With the commoditisation of the technology necessary for storing, computing, processing, analysing, and visualising big data, there is immense interest in leveraging such technologies for improving the efficiency of construction processes such as quality control (Bilal et al, 2016). Although the construction industry generates large amounts of data throughout the life cycle of a building, the adoption of big data technology in this sector lags behind the progress made in the aerospace and automotive industries.

In the report Imagining Construction's Digital Future (McKinsey, 2016), a Chapter considers digital collaboration, the Internet of Things (IoT) and advanced analytics. The Chapter mentions possibilities such as remote site inspections using pictures and tags (term assigned to a piece of information), radio-frequency identification (RFID) shared through mobile phone application software, plus the updating and tracking of live defect lists across projects. Though in the article there is much optimism associated with these technologies, in what is considered a fourth industrial revolution (PwC, 2016), in practice the training of the workforce is crucial for successful application on site.

Leading specifically onto quality assurance in 'Four Steps to Evolve Your Construction Quality Control Plan', preparation for digital construction methods can be overwhelming, but becomes much easier with the realisation that the whole progression is not activated simultaneously (Autodesk, 2018). The following stepped approach allows companies to make the suitable degree of change for their needs, and further evolve the processes:

- Implement a centralized document management system
- Bolster inspections tracking process with digital checklist inspections
- Correlate inspections to look-ahead programmes
- Perform collaborative inspections with trades and subcontractors

This approach provides a viable commencement in the application of digital technologies for quality assurance on site.

In a current major project in the United Kingdom, HS2 (High Speed 2) is committed to harnessing the power of digital engineering to build and operate a high-speed railway. Assets are designed, constructed and maintained digitally using both graphical and non-graphical input in a common data environment, providing real-time access to reliable and accurate information (HS2, 2017). This enables the exchange and management of co-ordinated quality digital data from multiple sources.

2.2.2 The Digital Engineering Environment

In 'Understanding the Implications of Digitisation and Automation' (Oesterreich and Teuteberg, 2016) the aim is exploring the current state of the art for the relevant technologies in the construction industry, with the following conclusions being drawn:

- The construction industry specific definition of the fourth industrial revolution (Industry 4.0) concept comprises a plethora of interdisciplinary technologies to enable the digitisation, automation and integration of the construction process at all stages of the construction value chain
- These technologies are on differing levels of maturity. Several technologies have reached market maturity and thus are currently available (BIM, cloud computing, mobile computing, modularisation). Whereas other technologies are still at the formative stage (additive manufacturing, virtual and mixed reality)
- Despite the given maturity and availability of many technologies, their widespread adoption by construction companies has yet to take place. However, there exists best practice which demonstrates the practical ways possible for the adoption of new technologies to digitalise and automate the construction process

The application of the technologies mentioned has great potential for use in quality assurance, being discussed in more detail within Chapter 2.2.3.

New relationships such as IoT enable a long-term partnership between the customer and the construction company that provide a building as a service. Changes to the service are planned collaboratively to adapt with the needs of the customer's business. In return, the construction company gets monthly revenue and a more predictable cash flow making their business planning less random (Woodhead et al, 2018). Though the paper particularly mentions sensors to gather information for quality records, many other processes are considered within this literature review Chapter.

At a company level, information and communication technology (ICT) is changing working practices. This has great implication for construction firms, providing a new challenge by reconfiguring their resources for the new technological processes on large construction projects (Redwood et al, 2017). ICT is changing three different aspects of large scale projects, as a tool during the build phase, for the ongoing building functionality (building management systems) and lastly benefitting occupant's experience. The integration of these three aspects are forcing directors to rethink their current ICT management systems. Project supply chains are being placed under contract in a reconfigured programme to take advantage of digital technology.

For senior managers implementing the systems an awareness of employee interaction with process change is required. This awareness is typified by the publications below:

- For staff to embrace a new habit that can involve curiosity and personal motivation, it is often considered that a period of around 66 days elapses (Lally et al, 2010), albeit the study was carried out in the wider field of social psychology as opposed to solely a construction workplace
- Though highlighted over three decades ago the sociological effects of information technology (IT) systems can still incur a bureaucratization of organisational structure, a shift in power to those in control of computer resources, although eventually employee job satisfaction does not appreciably decrease (Smith and Green, 1980)

- From the perspective of management input Mento et al (2002) provide models of the change process based on both theory and practice from exemplars in management literature. Also, Anderson (2018) reinforces the criticality of management involvement in the success of new IT integration
- Researchers in their efforts to clarify system use, develop tools for measuring computer user satisfaction. The Technology Acceptance Model (TAM) provides a basis for tracing the impact of external variables on internal beliefs, attitudes and intentions. This model suggests that perceived ease of operation and usefulness are the two most important factors in explaining process use (Legris et al, 2003). Though these two perceptions were deduced, a wide variance in the analysis was noted, implying the model can be further refined.

The paper 'Implementing Site BIM: a case study of ICT innovation on a large hospital project' discusses the practicality of implementation (Davies and Harty, 2013). Technical IT skills are adopted into the construction project through personal relationships and arrangements rather than formal processes. Execution is driven by construction project employees, rather than controlled centrally by the corporate IT function. This paper effectively reports on the reality of ICT when applied on a sizable complex project with co-ordinated design and construction occurring concurrently.

The manner a company handles its innovation platforms impacts on the successful implementation of digital engineering. At Crossrail for example, collaboration, culture and capability are the three key enablers that foster, nurture and incentivise innovation (Crossrail, 2013). The key elements for collaboration are partnering, funding, intellectual property and governance. For culture the elements are vision, leadership and strategic planning. Lastly for capability the elements are technical expertise, techniques, information systems for data, training and facilities. Though the intention is excellent, as the project is publicly funded, Crossrail is cautious on the extent of innovation to reduce the risk of failure from untested new technology.

In a review of software applications (apps) currently used by construction managers, most apps are not conceived for the construction industry (Azhar et al, 2015). Interviewees in this research also indicate that apps specifically related to quality control and deficiency tracking would be most useful. Also, the two main problems identified are the lack of training and difficulty viewing documents on mobile devices. It is noted that the operating system for the apps reviewed is predominantly Apple, though Android is now becoming more prevalent, being an open operating system where software additions can be more easily implemented.

2.2.3 Digital Engineering Quality Applications

For the use of digital engineering in quality management, IT in the form of document management systems is the enabler that links the various electronic processes together. The software of these systems contains an applicable programmable interface (API) and typically JavaScript object notation (JSON) file format. The standardisation of a product, service or process between supply chain members is assisted by PAS (Publicly Available Specification) 1192-2:2013 that is a specification for information management in construction using BIM (PAS, 2013). In practice the planning, design and construction of civil engineering projects are increasingly being managed and controlled through online collaborative platforms. Such cloud solutions provide the basis for a 'digital ecosystem' with potentially large productivity benefits (Cooper, 2018). When choosing document management systems for construction projects, effective electronic interfaces that stream data on demand are becoming available.

The following literature in Table 1 contains both quality assurance methods that begin addressing the limitations of existing processes and concepts that have been not widely implemented in business applications. Leading on from these research papers the current digital applications providing increased surety of quality outcome that are not broadly adopted to date are listed in Table 2, within the summary of this Chapter.

Table 1 Digital Engineering Quality Management Literature (2 pages)

Reference	Theme	Review	Topic Area
Chen and Kamara, 2011	"The paper introduces a framework for the implementation of mobile computing that comprises a technical model and an application model". The application model explores the likely interactions of "mobile computing, construction personnel and construction information"	Thorough analysis of possibilities that provide the way forward for digital engineering, though insufficient appreciation of education and training	Mobile phone and tablet
Leung et al, 2008	Presents a cost-effective construction site monitoring system integrating a long-range wireless network, network cameras and web-based collaborative platform	Includes a pilot study. As article published a decade ago available camera technology mentioned has become more sophisticated	Real time performance information, condition based monitoring
Hou et al, 2014	Context-aware mobile computing that integrates innovative concepts and technologies including but not limited to ICT, BIM, advanced visualisation and radio-frequency identification	"The integration of these technologies in a cohesive dynamically updated database combined with real time visualisation gives construction personnel a superior advantage compared to current practice"	ICT, BIM, terrestrial laser scanning, RFID, global positioning system (GPS), embedded tags
Wang and Chong, 2015	Research findings identify three phases where project performance can be improved by using BIM in pre-construction, construction and maintenance/operation phases	Maximisation of BIM requires full integration with other technologies across the project life cycle. Combines various technologies, studies carried out on live projects	BIM, Light Detection and Ranging (LiDAR), augmented reality, GPS, terrestrial laser scanning
Chen and Luo, 2014	Explores the advantages of 4D (3 dimensional and time) BIM for quality control by constructing a model in product, organisation and process (POP) data definition structure	High technology approach using 4D BIM (3D plus time element), this paper like many others does not include temporary works	4D BIM, POP

Park et al, 2013	Presents a conceptual system framework for construction defect management that integrates ontology and augmented reality with BIM	High technology approach to improve quality. At current stage of digital engineering use on site, could envisage being little employed	BIM, mixed reality
Lin et al, 2016	Approach integrates internet and BIM technologies to illustrate and analyse information at site in real time	Effectively facilitates quality inspections with identification and communication in a 3D BIM environment with the use of automated technology. Using BIM in this manner during the construction phase has significant potential in improving quality assurance	BIM, tablet, ICT
Lee et al, 2016	"Defect ontology is developed, work context information is taken from BIM models, extracted BIM data is converted to RDF" (resource description framework) "format and SPARQL" (standardised protocol and RDF query language) for resolution of queries	Highly technical approach from the perspective of current civil engineering projects, article openly admits that further in depth implementation is required for evaluating practical feasibility	BIM, photography, data collection template
Chang et al, 2013	Using 2D barcode and BIM technologies, this study proposes a BIM assisted quality inspection system	The article includes 2D bar codes, providing mobile platforms in localised areas and construction information for quality purposes. These 2D codes currently in use on sites, though arguably can now receive just the same information on a tablet with a data connection	BIM, 2D bar codes
Kwon et al, 2014	The paper develops systems firstly to enable quality inspections without visiting the real work site and secondly provides apps on tablets that allow personnel to automatically detect dimension errors and omissions on site	Though this study has been applied to reinforced concrete operations, the merging of these two processes can be successfully applied to range of construction operations	BIM, photography, internet, mixed reality
Kim et al, 2008	Objective of this research is a computerized quality inspection and defect management system which can collect defect data at site in real time	By collecting data in real time and immediately processing the output into an electronic quality	Mobile phone and tablet, wireless internet, ICT, real time performance information

Reference			
		management system, a construction team can focus on the issues without spending excessive effort manually producing records	
Akinci et al, 2006	This process utilises laser scanners for 3D spatial data and embedded sensors	Temperature sensors can measure gain of strength for a concrete pour. Laser scanners as discussed can verify an as-built model, though LiDAR is now popular for this process	Sensor, laser distance and ranging (LADAR)
Wang et al, 2015	Integrated system of BIM and LiDAR is provided for site quality information during collection and processing	System efficiently supports real-time quality control. Presently used on site by specialist company departments. Witnessed its effective use at railway depot in west London	BIM, LiDAR, real time performance information, point clouds of as-built construction
Wang, 2008	This study presents a novel and practical method to capture and represent construction project knowledge by using network knowledge maps	RFID tags track components for a project on and off site. Provides surety that specified item built into correct location aiding quality assurance. To encourage use could be a contract requirement	RFID, personal digital assistant

Reflecting on the literature for digital applications, a large group of articles, for example Wang and Chong (2015), Chen and Luo (2014), Lin et al (2016) and Chang et al (2013) concentrate on the use of BIM for digital quality assurance, though this process is only one of the potential options available. Whereas Park et al (2013), Lee et al (2016) and Kwon et al (2014) consider the use of digital applications to reduce defects that bears on the final quality, pursuing the goal of executing a task correctly at the initial attempt. Also, Kim et al (2008) and Akinci et al (2006) provide highly technical approaches, but construction personnel on site are currently unfamiliar with these techniques and yet to master the implementation. The papers by Chen and Kamara (2011) as listed in Table 1 plus Davies and Harty (2013) in Chapter 2.2.2 the Digital Engineering Environment both partly review the implementation of digital engineering, the subject of this research.

In conclusion, the literature reviewed, though covering various aspects digital processes without evident disagreement between articles, fails to address the coherent improvement of quality through the application of current digital technology. Consequently, the principal questions in this research focus on the delivery of a digitally connected approach to the management of construction quality assurance.

Chapter Summary

Table 1 contained both quality assurance methods that began addressing the limitations of existing processes and concepts that have been not widely implemented in business applications. This list reviewed digital frameworks, defect reduction, the involvement of BIM and detailed systems.

Within this summary, the following Table 2 lists possible digital quality assurance management processes for the building phase of civil engineering projects. Utilising this Table, by gaining advice from construction professionals in interviews, a quality management framework employing the most applicable digital technologies will evolve, as detailed in the subsequent two Chapters involving methodology and results.

Table 2 Possible Digital Quality Management Processes (2 pages)

Digital Process	Literature/Vendors	Advantages	Challenges
IT enabler	PAS, 2013	Main contractor and subcontractor project specification, quantities and drawings managed plus controlled through online collaborative platforms	Agreement of proprietary system before contract commencement
Mobile phone / tablet as personal digital assistant	Chen and Kamara, 2011; Kim et al, 2008; Hou et al, 2014	Provides the opportunity for a truly digital co-operative site quality assurance system	Education and training of all personnel so that the benefits can be appreciated
BIM	Chen and Luo, 2014; Lee et al, 2016; Lin et al, 2016	What is known as Level 3 BIM provides the integration of site activities into the project model, providing one depository of design, construction and operational information	The development of BIM for the construction phase across the industry in its early stages
Mobile and web application products	Aconex, 2018; Basestone, 2017; Viewpoint, 2018; Podfather, 2018	Supplied for construction teams to access, capture and communicate information. All field data for inspections and quality control can be shared via apps. Project delivery is collaboratively streamlined with the provision of reports, defects records and the flagging up of red line drawings (as-builts)	Ease of use and being robust in times of high data traffic are paramount
Digital inspection and test plan (ITP)	BIM360, 2018	Directly connects quality check sheets and resulting compliance documentation to a central document management system	In practice implementation to date very limited
Clash mitigation using BIM in construction phase	Wang and Chong, 2015; Akponeware and Adamu, 2017	Can analyse every possible clash from the many specialists involved, in a large project greatly enhancing quality and reducing rework	All designers, contractors and subcontractor using BIM in a common data environment
Real time performance information	Leung et al, 2008; Wang et al, 2015	Obtains evidence of component tolerance information/quality data during actual design or construction (real time performance information). The most impressive developments of quality assurance are in earthmoving operations (Navon, 2005)	Automated project performance control is a novel approach still in its infancy

Topic	Reference	Description	Comment
Mixed reality when digital BIM modelling and real content co-exist	Park et al, 2013; Kwon et al, 2014	Increased use of mixed reality when digital BIM modelling and real content co-exist, where architectural designers collaborate with construction teams transforming digital content into a physical object (University of Cambridge, 2017)	Expensive equipment that is vulnerable on site. Holograms useful in offices to elucidate building features
Point clouds of as-built construction	ClearEdge3d, 2017	Verity is a verification software that analyses point clouds of as-built construction using LiDAR against the design/fabrication model and flags up any out of tolerance that can be highlighted, annotated and electronically distributed	At present, generally being used only by specialists
Precise 3D vision to a mobile device	Structure, 2018	Structure Sensor adds precise 3D vision to a mobile device, giving it a new way of understanding the world around it. This enables a growing set of advanced capabilities like 3D scanning, indoor mapping and mixed reality experiences	Suitable for local use on a project though improved versions of the concept becoming available
Sensors within progressive manufacturing	Akinci et al, 2006	Sensors incorporated within the progressive construction process. For example, Utterberry 15 gram sensors operate to sub-millimetre precision, measuring multiple variables. These sensors collect, process and interpret data at source, transmitting information in real time (Chartered Institute of Building, 2014)	Likely to have only one use therefore becomes costly
Condition based monitoring	Pix 4D, 2017	Condition based monitoring, such as progress reports from photographs located on tower cranes	Provision of suitable vantage point can be demanding
3D geological models	Leapfrog 3D, 2017	Leapfrog 3D geological models that develop robust civil engineering groundwork solutions	Only as good as the information provided to the model
Character rigging and 3-dimensional animation	Mixamo, 2017	The software offers character rigging and three-dimensional animation that enables digital rehearsal of pedestrian regions and assists reviews of buildability	More use for method statements than direct quality applications
Business intelligence that reviews current media information	Batrinca and Treleaven, 2015	Business intelligence that reviews current media information applicable to a sector of an industry, in this case construction quality assurance	Value to be trialled on a live construction project

Radio Frequency Identification	Wang, 2008	Quality management systems which functions as a platform for gathering, filtering, managing, monitoring and sharing quality data technology	Cost currently significant factor limiting widespread use
Photogrammetric vision	Fathi et al, 2015	Fusion of photogrammetry and computer vision to support fully automated 3D modelling and mapping tasks, through the recordings of visual and electromagnetic sensors	Occlusion can occur whereby obstacles obstruct view between camera and object
Machine vision	Machine Vision, 2018	Machine vision and learning provides imaging-based automatic inspection for process control. System utilises industrial image processing with cameras mounted over production lines to visually inspect or direct products and guide robots	More suitable for factory off site prefabrications than on site construction
Simultaneous localization and mapping	Cadena et al, 2016	Technique used by generally autonomous vehicles to build up a map. The technique can formulate a map within an unknown environment or to update a map within a known environment. The output may involve the use of algorithms	Using autonomous vehicles in a site situation very challenging
Identification using 1-dimensional & 2D barcodes	Chang et al, 2013	Already in common use for mechanical and electrical plant identification, could be used much more extensively for structural components. Note a Quick Response (QR) Code is a type of matrix / two-dimensional (2D) barcode	Depends on having suitable site data connection from start on site
Backpack mobile mapping	Leica Pegasus, 2018	Backpack mobile mapping wearable reality capture platform that collects data indoors, outdoors and underground. Process could also be used by an unmanned aerial vehicle (drone)	Staff to be made aware of this service
Automated construction plant	Komatsu, 2018	Intelligent machine control, with parts of operations being completely automatic providing the possibility of improved quality control	Contractors still provide an operator due to safety considerations

Chapter 3. Method and Data Collection

This Chapter initially reviews the philosophical and methodological approaches of research, moving to design and associated methods that address the research's objectives and questions i.e. from Ontology and Epistemology, Methodology, Design, then to Methods. Consequently, the Chapter provides the method and data collection link between the digital processes available from the literature for quality assurance in Chapter 2 and the eventual processes chosen in Chapter 4. Please note the wording 'quality control' is used in the dissertation main title, as when the researcher has discussed the subject with lay persons, they are more comfortable with this terminology. Once into the detail of the dissertation the words 'quality assurance' become more predominant.

3.1 Philosophical Approach

Creswell (2014) provides frameworks for developing a research approach. Creswell refers to three components (philosophical worldview, research design and research methods) underpinning one of three research approaches, relating to quantitative, qualitative and mixed methods research. For this dissertation, the qualitative research uses a relativist approach, whereby the views of participants are utilised to discover the appropriate digital applications for construction phase quality assurance. Relativism considers that people have different ways of perceiving the world and there is no external reality independent of the beliefs and perceptions of those experiencing an event (Whitton, 2007). The complexity of experience and behaviour must be studied to gain true understanding.

It is noted that ontology is the philosophical supposition about the nature of reality while epistemology refers to the set of assumptions about the best ways to inquire into the nature of the world (Easterby-Smith et al, 2008). From this initial theoretical perspective, the ontology in this research is relativist, whereby the world is contextual and internal with investigation considering 'who', 'where', and 'when'. Within this relativist branch of ontology, the constructivist worldview for this research considers that reality is determined by the knower, matters are dependent upon human mental activity and structure relies on experiences or interpretations.

In the context of this dissertation the relativist strategy centres on researching the application of digital technologies in quality control for the build phase of civil engineering projects. The research is descriptive and statistical, providing data and characteristics about the population being studied.

3.2 Methodology

Methodology is the overall approach to the proposed research that links to the theoretical framework being applied. Building on the relativist ontology mentioned previously that seeks to establish meaning of a phenomenon from the views of participants, a qualitative methodology is established. This qualitative methodology is a means for exploring and understanding a social or human issue. Inductive data analysis builds from particular situations to themes. The methodology develops an initial understanding of an issue, with different perspectives between groups and categories of people.

The research approach uses 'grounded theory' considering processes based on the participants' views of the shortlisted digital technologies. Grounded theory is a design of inquiry from sociology in which the researcher derives a general abstract theory of a process grounded in the views of participants. Other genres of qualitative research such as described in Qualitative Data Analysis (Miles et al, 2014) are less suitable. For example, phenomenology tends to review data thematically to extract essences and essentials of participants' meanings. Also, ethnography stays closes to a naturalist form of enquiry that can involve extended contact within a given community. Also, Pickard (2013) notes that 'grounded theory' is an approach to research based on discovery, involving broad questions with focus emerging once data is observed and collected.

The inquiry involves using multiple stages of data collection, gained from literature plus interviews with the refinement and relationship of information categories (Creswell, 2014). This is followed by analysing the data to provide a quality management framework of the most applicable technologies. The search for suitable digital quality assurance processes utilises a worldwide literature review. The interviews seek the advice of construction professionals from varying specialist backgrounds.

There are three roles that the interviewer simultaneously fulfils (Knight and Ruddock, 2009). Firstly, the interviewer questions and probes the respondent, thus encouraging the meeting to be productive and thereby collecting beneficial data. Although many questions may be scripted, the qualitative interview affords an opportunity to explore issues not previously considered or that have emerged spontaneously during the interview. Secondly, the interviewer is listening to the respondent and evaluates their portion of the conversation. Thirdly, the interviewer also manages the process, including the timing and sequencing.

The following set of criteria defines the parameters of the technologies or processes that may be adopted and the sector of the construction industry being reviewed:

- Applicable to civil engineering projects of over £50 million expenditure in the United Kingdom, as sufficiently complex to warrant full application of digital quality engineering in its early stages and sufficiently knowledgeable personnel on site to appreciate the potential benefits
- Considers the application of digital engineering over the next 4 years in quality assurance during the construction phase of civil engineering infrastructure projects. This timescale due to rapid technological development enabling substantial evolution in the applications considered and the typical construction phase of a large civil engineering project involving a period of 4 years
- The research outcomes are acceptable to construction industry stakeholders who are principally designers (architects and structural engineers), the main contractors of an infrastructure project and their subcontractors. These companies reflect the existing contractual arrangements for project delivery
- The intended digital engineering applications to be employed on site have already been generally mentioned in academic articles, as this research focuses on amalgamating the practical use of various proposals
- The development of new digital applications from first principles is not attempted, though the merging of existing technologies is discussed

- The outcomes cover digital quality assurance from the perspective of a main project contractor or a client's project management team, as generally these operations are the promoters of digital engineering initiatives

3.3 Research Design

The research design is the plan of action that links philosophical assumptions to specific methods. The design is guided by the philosophical underpinning of the research, the aims and objectives, existing knowledge, the research questions plus available time and resources.

The overview of the design components is the claim, point of view, the research beliefs, the reasons or explanations, the evidence (support or grounds), data, examples that support the proposition, the researcher's argument and how the evidence provided leads to the claim.

The research gathers information by using a set of open-ended questions to stimulate the participants' answers (Flick, 2009). The data gathered is analysed to form themes or categories. The research looks for broad patterns, generalizations, or theories from these themes or categories.

Digital engineering is studied systematically and intervention is informed by theoretical considerations. This research is intended to foster a deeper understanding of the subject. The research is unique to this digital engineering study.

As the research is "real world", ethical research practices listed below were considered and undertaken whilst instigating the dissertation:
- Honesty
- Objectivity, striving to avoid bias results
- Integrity, keeping promises and agreements
- Carefulness, retaining effective records
- Openness with participants
- Respect for intellectual property, honouring patents and copyrights
- No plagiarism
- Confidentiality

The ethical research as mentioned (Booth et al, 2008), involves respecting sources, preserving data that runs against the results of findings, asserting claims only as strongly as warranted and recognizing the limits of certainty. A broader advantage occurs with the creation of a bond with the wider community centred on the benefits of research.

The interview research design is based on the following flow of actions:

- Formulate research questions
- Review literature
- Determine methods
- Consider appropriate population plus sample design, determine sample size
- Decide on mode of administration
- Develop and review questions, possible pilot interview and revise questions as necessary
- Administer method and transform into readable data
- Analyse data, interpret findings, and consider implications from research questions

In practice the research involves three stages:

Stage 1

A literature review discovers the full range of digital quality control concepts available in the construction industry to deduce a group of shortlisted technologies for quality management and the possible enabling environment required for successful implementation (Chapter 2).

Stage 2

By seeking advice regarding the merits of these shortlisted technologies through open ended questions during the interviews of the participants listed in Table 3, a digital quality management framework evolves (for results see Chapter 4).

Accordingly, for a project's build phase, through the purposeful selection of these participants, digital engineering quality processes within a connectivity enabled by technology can be enabled e.g. for continuing consultant design, contractor detailed design, off-site fabrication and on-site assembly.

The details of questions flow from the initial enabling environment necessary for digital technology, through to a general discussion on the future of the technology within construction and then the required specific applications.

Stage 3

Validation is carried out using the following techniques:

- Member checking, taking the near final themes back to participants to consider accuracy of interview reports compared to their statements
- Using a colleague, whose company role is Technical Assurance Manager, to critically appraise the dissertation, particularly methodology, methods and conclusions

It is noted that reliability is whether the results are repeatable, the ability to perform and maintain functions and deliver results consistently in defined circumstances (replication). Validity is obtaining results that accurately reflect the concept being measured.

This research design is a dynamic process therefore it becomes flexible. Early in the study links between applications, themes, previous findings and methods are postulated. It is intended that the links will form a coherent chain and may require adaption as the enquiry develops and findings emerge (Fellows and Lui, 2015). The goal is to maintain coherence and complementarity, providing robust results and conclusions.

3.4 Methods

Research methods are the various specific tools or ways data can be collected and analysed. The objective is to create empirical (primary) research data. Semi-structured interviews are used to narrow down topics, whereas completely un-structured interviews have the risk of not eliciting from the interviewees the themes more closely related to the research questions under consideration (Rabionet, 2011). Some specific questions require an answer, but at the same time the researcher wishes to discover the perspectives of the interviewees. Consequently, within the dialogue of the interviews, the format of an opening statement and a few general questions to prompt conversation is utilised.

This qualitative method of inquiry combines a pre-determined set of questions that encourage discussion with the opportunity for the interviewer to further explore evolving themes or responses. The same questions in the same order are asked to all interviewees to obtain quantifiable data on further analysis. In the development of the interviews, it was the intention that the questions should be open-ended, simple, specific, individual, exhaustive, neutral and balanced. Though this interview process is time and resource intensive, effective verbal communication is achieved.

Regarding sampling considerations from the whole population in this infrastructure sector of civil engineering, the sample has been selected on a non-probability basis. The sampling is not intended to be generalised (Martinez-Mesa et al, 2016), it is based on focused convenience (accessibility), part snowball characteristics (initial participants used to establish contacts with others), quotas of interviewees (intentional selection of specific number to meet established criteria) and purposive sampling (the opinion of professionals involved with digital engineering). Distortion in representativeness (prevention of bias) is minimised by the choice of varying disciplines for the interviewees, with knowledge being gained from the individuals and their respective companies across this infrastructure sector.

For particularly questions 3 (does the application of Building Information Modelling assist digitally connected quality assurance) and 5 (how do we ensure staff and subcontractors reap the potential benefit of the digital technologies), qualitative data analysis coding is carried out to seek structure to useful observations. This involved searching for themes, developing specific codes for types of participant responses and organising the resultant data. Accordingly, this provided a means to introduce interpretations into the research method. In detail, the coding progresses in stages to specific categories with due care to validity as described in Analysing Interview Transcripts in Qualitative Research (Burnard, 1991).

From a data cleansing perspective, excluding the dialogue unrelated to the thrust of the research, the transcribed original interviews remain unchanged with the initial open coding being freely generated. In practice a transcription and coding spreadsheet contains, for each question and each interview transcript, either the verbal response to a direct question or a coded output where the question seeks a perspective on an aspect of digital engineering. From this spreadsheet, a table is produced for the results of the questions. The full output from the direct question to discover the most appropriate digital quality technologies is placed in a separate table with the most popular being the processes chosen. After the presentation of the results from the interviews the data is reviewed for the findings obtained, the synthesis with literature and analysis of the outcome.

The interviews carried out are listed in the following Table 3. The interviewees are chosen to gain insight on quality issues from different perspectives of construction management including the design element. Assistance for the intended quality management framework is also obtained from subcontractors. These subcontractors carry out specialist works packages across the industry. Successful construction phase quality control rests with these companies adopting the main contractor's quality systems. Also, note that a Tier 1 main contractor is generally employed directly by a client, whereas Tier 2 are the subcontractors to the Tier 1 grouping.

This qualitative method is usually interested in identifying commonalities between types and then drawing out the implications of these commonalities to the larger whole. The emphasis on commonality means that once a qualitative researcher is conducting their interviews it may be found that that the evidence is so repetitive that there is no need to continue. Thus, saturation is central to qualitative sampling (Baker and Edwards, 2012). It was discovered for this research that saturation was achieved after 25 interviews.

Chapter Summary

After a discussion on the research approach, the method for this dissertation was explained including the wider issue of good practice. Fundamentally the method requires the choice of suitable digital applications to be gained from the responses of interviewees. Additionally, other contributory factors that impact on the application of digital engineering for quality assurance are considered in the interviews.

Table 3   **Interviewees**

Interview	Participant
01	Tier 1 Contractor, Head of Quality
02	Tier 2 Earthworks Contractor, Managing Director
03	Geosystems Business Partnership Manager
04	Design Management Advisor
05	Tier 2 Mechanical & Electrical Contractor, Design Manager
06	Tier 1 Contractor, BIM Lead Manager
07	Tier 2 Building Services Contractor, Communications Designer
08	Planning & Development Architect, Associate Director
09	Construction Engineering PhD Candidate
10	Tier 1 Contractor, Head of Digital Engineering
11	Tier 1 Contractor, Joint Venture Assurance Director
12	Major Client Senior Project Manager
13	Major Client Asset Manager
14	Tier 1 Contractor, Engineering Director
15	Consultant Project Manager
16	Tier 1 Contractor, Regional Design Manager
17	Tier 1 Contractor, Business Information & Tools Manager
18	Tier 1 Contractor, Design Package Manager
19	Tier 1 Contractor, Design Co-ordinator
20	Tier 1 Contractor, Digital Engineer
21	Tier 1 Contractor, Facilities Management Performance Director
22	Tier 1 Contractor, Senior Planning Engineer
23	Lecturer in Construction Project Management
24	Tier 1 Contractor, Project Manager
25	Tier 1 Contractor, Senior Engineer

Chapter 4. Results and Discussion

The review of current commercial technologies available in the construction sector that could be greater utilised in quality management was summarised in Table 2 during the latter part of Chapter 2. Using the methodology described in the preceding Chapter 3, the assessment process provides responses to seven questions, including the digital technologies that have the greatest potential to improve quality management. This Chapter provides the results from the interviews, the findings, the outcome analysis and synthesis with the literature review.

4.1 Results

With reference to Table 4, considering each question in sequence:

For question 1, the principal documents that bind the quality assurance processes together are the project management plans followed equally by Publicly Available Specification 1192-2 (PAS 2013), digital engineering management plans and digital inspection & test plans. Participants from Tier 2 subcontractors appreciate that Tier 1 main contractors decide the management plans for a particular project.

For question 2, the changes required to existing processes to enable fully digital quality applications are co-ordinated procedures followed by a suitable IT enabler. Also, Tier 1 contractors should work collaboratively with vendors to provide systems in a common data environment. Preferred digital quality processes are to be developed with further trials on site. Systems that can be easily transferred to fresh projects are to be encouraged. Three respondents considered projects are to be provided with a digital engineer to set up and promote digital applications.

For question 3, ten interviewees stated the application of Building Information Modelling reasonably assists digitally connected quality assurance. Interviewees were conscious that an awareness of BIM encouraged the use of other digital applications.

For question 4, in the list of digital technologies indicated at the interviews in Table 2, all participants considered that there were no unaccounted processes but there were additional applications within the processes. Sensors for example, besides their use in the initial civil engineering fabric of a structure can be used to extensively monitor the mechanical and electrical aspects of an infrastructure asset.

Table 4 Principal Interview Results

	Questions 1 to 7 - Principal Results		
1	What are the principal documents that bind the quality assurance processes together?	Management Plans followed equally by PAS, specific digital engineering plans and ITP's	
2	What are the changes required to existing processes in your preferred document management systems to enable fully digital quality applications?	Co-ordinated processes followed by suitable IT enabler	
3	Does the application of BIM assist digitally connected quality assurance? Welcome to expand your response	Greatly	7
		Reasonably	10
		Little	4
		Not at all	4
4	In the list of digital technologies indicated at the interview do you believe there are any missing applications appropriate for construction phase quality management?	No further applications to be added	Listed processes remain the same
		Additional	Additional applications within processes
5	How do we ensure staff and subcontractors reap the potential benefit of the digital technologies?	Education	Mentioned 2 times in interviews
		Training	14
		Organisation	13
		Culture	7
		Commonality	2
		Witness	5
		Incentivise	1
6	Your thoughts please for the future of digitally enabled quality assurance in the construction industry?	See Chapter 4.2 Discussion, though it is noted that all interviewees considered digital engineering part of the future for quality assurance	
7	Which suite of these digital technologies indicated at the interview is the most appropriate for the next four years?	For details of results see Table of chosen Digital Engineering Applications for Quality Assurance in Appendix D	

For question 5, to ensure staff and subcontractors reap the potential benefit of the digital technologies, interviewees reckoned training followed by organisation were the key enablers. Also, it was viewed that training is a learning process during employment, whilst education occurs at school and organisations such as universities.

For question 6, participants' views on the future of digitally enabled quality assurance in the construction industry are fully reviewed in the discussion section of this Chapter and in Chapter 5 Implications. All, including the more facilities management based personnel, considered that digital engineering is part of the future for quality assurance.

For question 7, the suite of these digital technologies indicated at the interview most appropriate for the next four years is listed in the following Table 5.

Table 5 Framework of Technologies for Digital Quality Management

Digital Engineering Application	Wide Use	Localised Implementation
Information technology enabler	✓	
Mobile phone / tablet as personal digital assistant	✓	
Building Information Modelling (BIM)	✓	
Mobile and web application products	✓	
Digital inspection & test plan		✓
Clash mitigation using BIM in construction phase	✓	
Real time performance information	✓	
Point clouds of as-built construction	✓	
Precise 3D vision to a mobile device	✓	
Condition based monitoring		✓
3D geological models		✓
Radio Frequency Identification		✓
Photogrammetric vision		✓
2D (QR code) and 1-dimensional barcodes	✓	
Backpack mobile mapping		✓

In Table 5, 'wide use' is considered applications applied in projects by more than 75% of participants and 'localised implementation' by more than 50% of participants.

The research aim to undertake a comprehensive explorative analysis of the potential improvement of quality control in the construction industry with the use of digital technology has been achieved using information obtained from the Chapter 2 Literature Review and the broad perspectives of the 25 interviewees. This approach has enabled the appropriate applications to be discovered.

The dissertation's objective of current commercial technologies available in the construction sector that could be greater utilised in quality management have been examined in Chapter 2. An assessment process to determine what technologies have the greatest potential to improve quality management was established in Chapter 3. The changes required in processes within quality management systems to accommodate this technology are evaluated from Question 2 in the interviews.

The original research questions stated in Chapter 1.4 of the Introduction were incorporated into the interview questions in Appendix A. The consequent provision of the framework of technologies for digital quality management in civil engineering from the analysis of the interviews is based on the rich seam of information articulated in the 25 interviews, then transcribed with the principal informative text set out in a spreadsheet. The findings from the spreadsheet are stated in this Chapter along with the outcome analysis.

4.2 Discussion

Having listed the digital technologies that have the greatest potential to improve quality management and a summary of the responses from the interviewees, the following discussion provides a detailed outcome analysis.

The findings of the research are as follows:

1) These digital quality processes have been available for many years, with literature on the subject or their individual references in Table 1 extending back often a decade

2) Apart from document management systems, currently the regular applications on site are quite limited, albeit specialist company departments employ many of the possible technologies

3) In the opinion of the interviewees 68% of the possible applications listed in Appendix B will be applied within the next four years on major civil engineering projects

4) Practical applications for site use dependent on developments by vendors

5) Processes being witnessed in use at project level enables personnel to appreciate the benefits of digital applications

The analysis of these findings leads to the outcomes listed below:

1) The project management plans followed equally by PAS, specific digital engineering plans and ITP's are typically the core procedures that act as the starting points to bind the quality assurance data together

2) To be visible in an electronic format with a connectivity enabled by technology, the data is enabled by a document management system

3) Principal output should be viewable on mobile devices with full details on laptops. For site activities where data connectivity is poor or unavailable, paper records need to be easily merged into electronic systems

4) Interface issues exist between systems that current common data platforms only partly resolve

5) The training and encouragement of site teams in the use of digital technologies is as important as the actual applications themselves

6) Though the value of the individual applications is beneficial, the real advantage is utilising the data provided in a holistic approach across the spectrum of digital engineering

7) BIM is one part of this holistic approach. Its use stimulates the application of other digital processes

8) Vendors provide products in a simpler format than implied by the academic papers in Table 2

In the synthesis with the literature reviewed, Chen and Kamara (2011) in the framework for the implementation of mobile computing on construction sites, correctly identify three major components: mobile computers, wireless networks and mobile applications. In the use of real-time communications, Leung et al (2008) approach the improvement of quality by effectively monitoring projects via cameras, providing progress data so that the supply chain can communicate remotely on issues which if unresolved result in defects. In Table 2 of Possible Digital Quality Management Processes, certainly condition based monitoring utilising cameras on tower cranes is following this approach. Hou et al (2014) in the use of digital technologies in productivity improvement take advantage of ICT, RFID, smart tags, laser scanning, BIM and augmented reality. Within this journal paper it is acknowledged these processes also improve quality and again are listed in Table 2. Also these processes are supported by interviewees 10, 17 and 20 who are all directly involved with digital technology. Wang and Chong (2015) state that BIM in the construction phase can improve project performance though must be integrated with other technologies. The participants in Table 3 concur with this statement. Chen and Luo (2014) propose a BIM based quality management model, that to be effective, requires direct field data transfer to the model. At present, this generally does not occur and is unlikely to happen with the 4-year timeframe for application of digital processes in this research. Park et al (2013) propose a construction defect management framework involving a ontology-based data collection template. Ontology represents knowledge as a set of concepts within a domain (in this case defects) and the inter-relationships between those concepts. Albeit a quite credible process, at this stage of digital application use on site, this framework is an item for future development. Kim et al (2008) employ personal digital assistants and a wireless web-integrated system for quality inspections. This system is favoured by all participants in this dissertation. Wang et al (2015) propose integrating BIM and LiDAR for real-time quality control and Wang (2008) propose enhancing construction quality inspection and management using RFID technology. Both these proposals are included in Table 2 and receive favourable responses from the interviewees.

Within the Chapter 2 Literature Review, beside the more technical papers just mentioned, during the interviews the importance of understanding the digital engineering environment for the implementation of the processes leads back to the articles in Chapter 2.2.2. For this understanding Oesterreich and Teuteberg (2016) plus Redwood et al (2017) provide a good starting point. Participants concurred with the views of Anderson (2018) that management involvement for the success of new IT integration is imperative.

35

Considering interview questions 1 to 6 in more detail, for question 1, ITP's and specific digital engineering management plans, though generally considered less important than the overall project management plan, are nevertheless viewed paramount by 24% of participants. Certainly, a good ITP can link the CAD drawings to the proof of a defined digital quality standard within the client's sectional completion certificates. As well digital engineering plans concentrate a project's positive efforts on this relatively new digital approach. Additionally, the management process and its visualisation should replicate the perception of reality for site personnel out in the field, with a feedback loop to improve the process. One participant from the rail industry (interviewee 13) noted a very simple quality operational success story for a project is achieving the list of client 'functional requirements' which on occasions are listed in the contract documents.

For question 2, 16% of participants stated the common data environment is the initial key enabler for the changes to existing processes. To provide this environment across the broad spectrum of clients, designers, main contractors and subcontractors remains a great management challenge, especially when major operations each have their own preferred data management system, initiated with all good intentions but prevent commonality.

For question 3, 32% of interviewees asserted that currently BIM has little or no impact on site in assisting digital engineering applications for quality assurance. This attitude is based on the present project situation, whereby although electronic processes are now being utilised for quality purposes, many personnel are unaware of BIM being applied in the construction phase.

For question 4, though it was considered Table 2 successfully covers the list of digital processes, 40% of participants mentioned the evolvement of applications within the groupings. Even during the course of this dissertation, it has become evident that much development is being brought to the market place by vendors in applications such as geographic information systems (GIS). As well the gains from the individual applications can be leveraged, for example by mounting photogrammetric vision apparatus on drones.

For question 5, to ensure a site team including its supply chain reap the benefits of what could be considered a digital twin providing quality data, 28% of

36

participants mentioned the importance of the project culture of both the client and principal contractor. To realise a digital engineering culture both behavioural change and the willingness of senior management to innovate were mentioned as essential. Other factors acknowledged as being significant were witnessing use of digital applications on site, commonality of applications across projects led by main contractors, possibly incentivising use so the spread of the technology across the industry is faster and digital engineering forums as part of organisations such as the Civil Engineering Contractors Association.

For question 6, considering the future of this digital world of quality assurance, a sobering response from interviewee 11 in Table 3 was that it will not necessarily improve the excellence of a structure, as an organisation still must know what is being built. Also, interviewee 25 reckoned the construction industry is moving towards complex digitally engineered management but lack the trades personnel to execute activities correctly, resulting in very sophisticated computed based defect sheets, when really subcontractors should be providing correct components at the first attempt. Interviewee 21, embracing an optimistic scenario, summed the future of digital technology very succinctly. The participant stated quality in the construction process will improve due to more effective engagement of the groups involved in the design and construction of the structure. These groups include planning teams, cross-functional production management teams, the supply chain, and trades persons on site. This ability to engage in a highly visual manner enables teams to discover issues ahead of construction, plus anticipate these issues from data and learning, preventing defects occurring by exchanging expertise.

For a critique of the results, by reflecting on the transcribed data of the participants' conversions, the following is noted:

- Overall impression arises that the on-going process of digitalisation for quality records is irreversible
- Uptake of these digital quality assurance processes on major projects has notably increased even during the evolution of this research
- Successful implementation at project level depends on the management of change, as much as the applications themselves

- Unfortunately, an excellent digital quality assurance process may still involve much rework, if the capabilities of the workforce in the subcontract packages deteriorates
- Participants cautious on the positive outcomes, as concerns exist about other aspects of the industry, such as competitive tendering encouraging under-resourced contractors and insufficient fully agreed design information at time of construction

Also, these five findings confirm the viewpoint of the researcher, during a working life of four decades in construction management.

To conclude for this Chapter, quality assurance is required due to client expectations, plus for the reduction defects and any consequent rework. It is considered that assurance can be improved with the use of digital engineering. By selecting processes from this research the benefits of digital technology can be applied, due to the assistance of interviewees deducing co-ordinated applications. This results in a holistic approach of applying smart technologies, hence success in utilising digital engineering. Therefore, the provision of enhanced quality assurance is achieved. This connected approach to quality assurance is delivered through harnessing the specialist technologies and the capabilities of business intelligence, BIM, surveying, GIS, automation, collaboration and digital mobility.

Chapter Summary

In Table 4 the principal interview results were provided, with the framework of technologies for digital quality management in Table 5. The following discussion provided analysis of the interviewees' responses and the literature review. The most significant outcome is that the applications are readily available once the principles are understood by company management. The importance of this outcome is that the implementation becomes an issue of training and company willingness at director level for the adoption of digital engineering.

Chapter 5. Implications

This Chapter provides the implications from the output of the research that was explained in the previous Chapter. These implications are substantially drawn from participants' responses to interview questions. The Chapter contains sections for practice and academia. The bracketed numbers refer to the source of the implications from the interviewees listed in Table 3.

5.1 Implications for Practice

The implications for practice are as follows:

1) Varity of individual digitals applications that have matured over many years are available from vendors. Currently all these different systems on various electronic platforms are not necessarily compatible with each other. Tier 1 contractors are moving in the direction where everything is visible in one platform (15)

2) The evolution of data consumption is leading towards streaming the latest information received (10). The way IT is set up and managed is what makes the digital applications useful (05), though costs must be considered (02)

3) Once a common database is achieved, with Tier 1 & 2 contractor IT requirements in the conditions of contract (17), then personnel can understand how processes are performing (21). Also, broken chains where clients and contractors are using different platforms are prevented (03)

4) Within this linked database approach sits the improvement of quality assurance on a project by digital methods. A robust system that ensures a successful step by step operation can be created (04). Including specific digital processes, in the subcontract requirements for the supply chain, aids instigation (07)

5) As digital applications at an early stage of implementation on site, initially important for personnel to become familiar with using tablets and mobile phones to access information (16)

6) Place personnel psychologically in charge of the technology so there is a sense of owning the process (23)

7) Digital quality control utilising GIS is very close to fruition (09). It is considered that surveying processes will incur the largest change within the next decade (03)

8) 3D approach to drawings will become the norm (05)

9) Clients are beginning to understand the benefit of being able to digitally track quality (08), albeit this requires capital investment (12)

10) A concern exists that will all enterprises willingly take on board these digital processes or will they have to be coerced into adopting them (10), especially to gain the benefits that accrue from all parties co-operating across a common platform (07)

5.2 Implications for Academia

The implications for academia as follows:

1) For the application of digital processes, competency of use is paramount (13)

2) Consequently, suitable education at school and university is required (02, 10)

3) The above occurs alongside company training particularly at a project's commencement to bring both employees and the supply chain on board (11, 20)

4) IT appears very persuasive on a computer screen, so personnel must be very capable in providing resilient quality information (4)

5) Now much development of applications by vendors (18). A beneficial feedback loop can be created through to academia

6) For the academic papers quoted in Table 1, 57% refer to BIM enabled processes of quality assurance in the construction phase. In practice this little occurs at present, especially in comparison with the frequent application of BIM in the design phase. Fully integrating site quality assurance data within a BIM design model could be quite transformational (01)

7) For effective utilisation in the construction phase, the I (information) in BIM needs to be readily available besides the model (17). Any study that enhances this information aspect would be very advantageous

8) Integration of these digital technologies across the wide supply chain of an infrastructure project presents a great challenge (14). For example, in one new junction on the M20 motorway there are 50 subcontractor packages (25)

9) To implement digital engineering, a technologically exciting environment for encouraging people to join the construction industry should be created, even initially during the education phase (24)

10) In design, shortly for quality assurance and already in facilities management, digital applications are becoming very prevalent. So, research into construction phase automation leads onto the final part of the building process that to date been unaffected by the recent march of technology (23)

Chapter Summary

As a consequence of the interviews, it is evident that on-site implementation of digital engineering for quality assurance is at an early phase of development. There are elements of digital technology, such as earthworks incorporating geo-spatial setting out, that certainly deliver improved site quality control. For both main contractors and subcontractors the principal goal presently is for staff and operatives to become used to the world of mobile technology, using tablets and smart mobile phones to record and transmit quality assurance data.

Many items in Table 2 that provides possible digital quality management processes are currently only used by specialist departments. Also, the application of digital engineering for quality matters is not only a choice of systems, but the suitability of IT infrastructure and the body of technical knowledge available in the industry.

Chapter 6. Conclusions

6.1 Summary

The research questions stated in the Introduction were incorporated into the interview questions, with the summary of finding, from the rich seam of information articulated in the 25 interviews, provided in this Chapter. This approach has enabled the research aim of undertaking a comprehensive explorative analysis of the potential improvement of quality control in the construction industry with the use of digital technology to be achieved, using information obtained from both the Chapter 2 Literature Review and the participants' perspectives.

Digital quality applications that will be in wide use within the next 4 years:

- Information technology enablers such as electronic document management systems
- Mobile phone and/or tablet as a personal digital assistant
- Building Information Modelling
- Mobile and web application products
- Clash mitigation using BIM in the construction phase
- Real time performance information
- Point clouds of as-built construction
- Precise 3-dimensional vision to a mobile device
- Both 2-dimensional (QR codes) and 1-dimensional barcodes

Applications that will be less widely implemented:

- Digital inspection and test plans
- Condition based monitoring
- 3-dimensional geological models
- Radio Frequency Identification
- Photogrammetric vision
- Backpack mobile mapping

For major projects in the construction industry, quality assurance is required due to client expectations, plus to reduce defects and consequent rework. It is considered by all participants in this dissertation's interviews that quality assurance can be improved with the use of digital engineering.

This digital engineering technology is a holistic and collaborative approach that enables personnel to make smart decisions with data. The connected tactic to better information management is delivered through harnessing the specialist technologies and capabilities of business intelligence, BIM, surveying, geographic information systems, automation, collaboration and digital mobility. Within this linked approach sits the improvement of quality assurance on a project by digital methods. Note BIM is only one component of this improvement process.

These digital methods for quality assurance are yet to be fully provided in a common data environment. As well, implementation of digital engineering for site quality assurance is at an early stage of development. In the initial list of digital applications of this research (Table 2) for possible instigation on site, most applications are currently only used by specialist departments. Albeit a variety of individual digital quality applications that have matured over many years are available from vendors. The technology operates quite satisfactorily, applying the process is now the issue.

There are construction processes such as earthworks, incorporating geo-spatial setting out, already providing improved quality control utilising digital technology. For both main contractors and subcontractors, the primary goal is presently for staff and operatives becoming familiar with the world of this technology, then using smartphones or tablets to record and transmit quality data. Training, particularly at a project's commencement, is required to bring both employees and the supply chain on board.

The application of digital engineering to quality issues is not only a choice of systems, but also the suitability of IT infrastructure and the body of technical knowledge available in the construction industry. The evolution of data consumption is leading is towards streaming the latest information received. The way IT is set up and managed is the enabler making digital applications useful. Also, once a common database is achieved within Tier 1 and 2 contractor IT outputs and specified in the conditions of contract, then the overall project team can readily understand how quality processes are performing. Consequently, broken chains, where clients and contractors are using different electronic platforms cumulating in poor communication, are prevented.

The most significant outcome is the applications are readily available once the principles are understood by company management. The importance of this outcome is implementation then becomes an issue of training and company preparedness for the adoption of digital engineering.

For successful application, company willingness involving senior management is necessary. It should be appreciated that integration of these digital technologies across the wide supply chain of an infrastructure project presents a great challenge.

6.2 Limitations of this Research

The research criteria principally involve civil engineering projects of over £50 million expenditure in the United Kingdom during the next 4 years. This scale of project currently warrants the investment in the full spectrum of digital engineering, though over the coming years the use of digital processes will become much more widespread, as the awareness of the benefits is more widely dispersed throughout the industry.

A small sample size was undertaken in a geographical sense. There was insufficient time available to have more participants, though each hour-long interview provided a good source of data. A more detailed and in-depth analysis of the subsequent findings was again limited by the period available, consequently further recommendations are stated in this Chapter.

6.3 Contributions to Knowledge

The expansion of knowledge commences with four key contributions:
1) Awareness of digital quality technology
2) An explanation of the digital quality processes
3) Senior management and project teams benefit from particularly the research summary when setting up digital quality assurance systems
4) Enabling the journey to digital engineering in the construction phase that may progress to greater automation of building activities

For working practice in the industry, to provide an awareness and explanation of digital quality assurance, Table 1 and particularly Table 2 are effective starting points. Table 1 contains both quality assurance methods that start addressing the limitations of existing processes and concepts that have been not widely implemented in business. This list reviews digital frameworks, defect reduction, the involvement of BIM and detailed systems. From Table 2, listing possible digital quality assurance management processes, a quality management framework employing the most applicable digital technologies evolves. This framework develops from the research carried out as detailed in the two Chapters involving methodology and results.

For academia, enabling the journey to digital engineering in the construction phase, competency of use is paramount. Consequently, suitable education both at school and university is required. This links with company training for both employees and the supply chain during the construction phase of a project. A beneficial feedback loop from developments of applications by vendors can be created through to academia.

6.4 Recommendations

The configuration of procurement timing with programme requirements of the digital processes to be instigated, as subcontractors providing specialist design elements require a sufficient period to align their IT, so their systems are appropriate for digital quality assurance in the construction phase. Tier 1 contractors should work collaboratively with subcontractors, vendors and academia to provide systems in a common data environment.

Preferred digital quality processes are to be developed with further trials on site. As well systems that can be easily transferred to fresh projects are to be encouraged. It would be beneficial to engage with other projects, throughout the United Kingdom initially, to obtain a more uniformed consensus. A useful recommendation is projects being provided with a digital engineer to set up and promote digital quality applications.

The review and then enhancement of current education and training for digital engineering assists the successful promotion of the technology. For effective utilisation in the construction phase, the I (information) in BIM needs to be readily available. Any study that enhances this information aspect would be very helpful.

The costs of implementing these digital technologies is outside the scope of this research. It is suggested that a cost and potential benefit analysis should be executed, including the review of initial set up arrangements and the possible reduction of defects in monetary terms.

To conclude digital applications such as 3D models are already prevalent for design. Shortly the growing use of digital technology for quality assurance in civil engineering infrastructure projects will take place, as indicated by the views of this dissertation's participants. It was also learnt that facilities management operations are increasingly using digital systems. So, increasing digital engineering research for the actual building works, encourages the automation for the outstanding element of the construction process that to date been unaffected by the recent march of automated technology in other industries.

REFERENCES

Aconex, 2018. *Cloud platform connecting teams on the world's largest infrastructure projects.* [online] Available at: <https://www.aconex.com/> [Accessed 10 February 2018]

Akinci, B., Boukamp, F., Gordon, C., Huber, D., Lyons, C. and Park, K., 2006. *Formalism for Utilization of Sensor Systems and integrated project models for active construction quality control.* Automation in Construction Vol 15, pp 124-138

Akponeware, A. and Adamu, Z., 2017. *Clash Detection or Clash Avoidance? An Investigation into Coordination Problems in 3D BIM.* MDPI Buildings Vol 8, Issue 3, Article 75

Anderson, T., 2018. *Understanding the Success or Failure of Organizational ICT Integration: The Criticality of Managerial Involvement.* [online] Available at: <https://www.tandfonline.com/doi/full/10.1080/14697017.2018.1491482> [Accessed 21 September 2018]

Autodesk, 2018. *Construction Quality Control Plan.* [online] Available at: <https://connect.bim360.autodesk.com/construction-quality-control-plan> [Accessed 22 January 2018]

Azhar, S., Jackson, A., and Sattineni, A., 2015. *Construction Apps: A Critical Review and Analysis.* McWhorter School of Building Science, Auburn University, Alabama, USA

Baker, S. and Edwards, R., 2012. *How many qualitative interviews is enough.* [online] Available at: <http://eprints.ncrm.ac.uk/2273/> [Accessed 24 March 2018]

Basestone, 2017. *Basestone digital delivery platform for construction.* [online] Available at: <https://basestone.io/> [Accessed 10 February 2018]

Batrinca, B. and Treleaven, P., 2015. *Social Media Analytics: a survey of techniques, tools and platforms.* AI & Society, Springer Vol 30, pp 89-116

Bilal, M., Oyedele, L., Qadir, J., Munir, K., Ajayi, S., Akinade, O., Owolabi, H., Alaka, H. and Pasha, M., 2016. *Big Data in the construction industry; a review of present status, opportunities and future trends.* Advanced Engineering Informatics Vol 30, pp 500-521

BIM360, 2018. *Quality Management.* [online] Available at: <https://bim360.autodesk.com/construction-management-software/quality-management> [Accessed 17 February 2018]

Booth, W., Colomb, G. and Williams, J., 2008. *The Craft of Research.* The University of Chicago Press, Chicago

Broyd, T., 2017. *Institution of Civil Engineers Presidential Address: Engineering a Digital Future.* Civil Engineering Vol 170 Issue 1, pp 3-8

Bryde, D., Broquetas, M. and Volm, J., 2013. *The Project Benefits of Building Information Modelling (BIM).* International Journal of Project Management Vol 31, pp 971-980

BSI, 2017. *Little Book of BIM.* [online] Available at: <http://pages.bsigroup.com/l/43652/2017-08-22/jwsk2l?utm_source=BSI-Website&utm_medium=online&utm_content=LBB&utm_campaign=BP3> [Accessed 20 November 2017]

Burnard, P., 1991. *A method of analysing interview transcripts in qualitative research.* Nurse Education Today Vol 11, pp 461-466

Cadena, C., Carlone, L., Carrillo, H., Latif, Y., Scaramuzza, D., Neira, J., Reid, I. and Leonard, J., 2016. *Past, Present and Future of Simultaneous Localization and Mapping: Towards the Robust-Perception Age.* Institute of Electrical and Electronics Engineers, Transactions on Robotics 32(6), pp 1309-1332

Chan, P., 2011. *Quality Assurance in the Construction Industry.* [online] Available at: <http://www.tandfonline.com/doi/abs/10.1080/00038628.1996.9697365> [Accessed 15 February 2018]

Chang, J., Su, Y. and Lin, Y., 2013. *Development of Mobile BIM assisted Defect Management System for Quality Inspection of Building Projects.* 13th East Asia-Pacific Conference on Structural Engineering and Construction, Sapporo, Japan

Chartered Institute of Building, 2014. *Digital Innovation Award: Utterberry Wireless Sensors for Built Environment Monitoring.* [online] Available at: <http://iandrawards.ciob.org/node/66> [Accessed 14 November 2017]

Cheetham, D. and Carter, D., 1993. *The challenge of assuring quality on site.* Building Research and Information Vol 21 No. 2, pp 85-98

Chen, L. and Luo, H., 2014. *BIM-based Construction Quality Management model and its Applications.* Automation in Construction Vol 46, pp 64-73

Chen, Y. and Kamara, J., 2011. *A framework for using mobile computing for information management on construction sites.* Automation in Construction Vol 20, pp 776-788

ClearEdge3d, 2017. *Verity Construction Verification Software.* [online] Available at: <http://www.clearedge3d.com/verity-construction-verification-software/> [Accessed 11 November 2017]

Cooper, S., 2018. *Civil engineering collaborative digital platforms underpin the creation of digital ecosystems.* Institution of Civil Engineering Proceedings Vol 171, Issue CE1, p 14

Creswell, J., 2014. *Research design: Qualitative, quantitative, and mixed methods approaches.* Sage Publications Ltd., London

Crossrail, 2013. *Crossrail Learning Legacy: innovation platform.* [online] Available at: <https://learninglegacy.crossrail.co.uk/documents/innovation-strategy/> [Accessed 18 November 2017]

Crossrail, 2016. *Crossrail Learning Legacy: quality dashboard.* [online] Available at: <https://learninglegacy.crossrail.co.uk/documents/quality-dashboard/> [Accessed 20 November 2017]

Dale, B., 2006. *Managing Quality.* Blackwell Publishing, Oxford, England

Davies, D. and Harty, C., 2013. *Implementing 'Site BIM': a case study of ICT innovation on a large hospital project.* Automation in Construction Vol 30, pp 15-24

Davila Delgado, M., Butler, L., Gibbons, N., Brilakis, I., Elshafie, M. and Middleton, C., 2017. *Management of structural monitoring data of bridges using BIM.* Institution of Civil Engineering Proceedings Vol 170 Issue 3, pp 204-218

Easterby-Smith, M., Thorpe, R. and Jackson, P., 2008. *Management Research.* 3rd edition, Sage Publications Ltd., London

Fathi, H., Dai, F. and Lourakis, M., 2015. *Automated as-built 3D reconstruction of civil infrastructure using computer vision.* Advanced Engineering Informatics Vol 29, pp 149-161

Fellows, R. and Lui, A., 2015. Re*search Methods for Construction.* John Wiley & Sons, Chichester, United Kingdom

Flick, U., 2009. *An Introduction to Qualitative Research.* 4th edition, Sage Publications Ltd., London

Hou, L., Wang, X., Wang, J. and Truijens, M., 2014. *Integration Framework of Advanced Technologies for Productivity Improvement for LNG Mega-Projects.* Journal of Information Technology in Construction Vol 19, pp 360

HS2, 2017. *What-does BIM mean for HS2*. [online] Available at: <http://www.rsgbestpractice.org/wp-content/uploads/2016/08/what-does-bim-mean-for-hs2.pdf> [Accessed 9 November 2017]

Kim, Y., Oh, S., Cho, Y. and Seo, J., 2008. *PDA and wireless web-integrated system for quality inspection and defect management of apartment housing projects*. Automation in Construction Vol 17, pp 163-179

Knight, A. and Ruddock, L., 2009. *Advanced Research Methods in the Built Environment*. John Wiley & Sons, Chichester, United Kingdom

Komatsu, 2018. *Intelligent Machine Control*. [online] Available at: <https://www.komatsu.eu/en/Komatsu-Intelligent-Machine-Control> [Accessed 10 March 2018]

Kwon, O., Park, C. and Lim, C., 2014. *Defect Management System for reinforced concrete work utilising BIM, image-matching and augmented reality*. Automation in Construction Vol 46, pp 74-81

Laing O'Rourke, 2018. *Engineering Excellence Journal edition 2* [online] Available at: <http://www.laingorourke.com/engineering-the-future/digital-engineering-2/eej/> [Accessed 20 January 2018]

Lally, P., Van Jaarsveld, C., Potts, H. and Wardle, J., 2010. *How are habits formed: Modelling habit formation in the real world*. European Journal of Social Psychology Vol 40, pp 998-1009

Leapfrog 3D, 2017. *Leapfrog 3D Geological Models*. [online] Available at: <http://www.leapfrog3d.com/industry-solutions/civil-and-environmental> [Accessed 9 November 2017]

Lee, D., Chi, H., Wang, J., Wang, X. and Park, C., 2016. *A linked data system framework for sharing construction defect information using ontologies and BIM environments*. Automation in Construction Vol 68, pp 102-113

Legris, P., Ingham, J. and Collerette, P., 2003. *Why do people use information technology? A critical review of the technology acceptance model*. Information & Management Vol 40, pp 191-204

Leica Pegasus, 2018. *Backpack Wearable Mobile Mapping*. [online] Available at: <https://leica-geosystems.com/en-gb/products/mobile-sensor-platforms/capture-platforms/leica-pegasus-backpack> [Accessed 14 February 2018]

Leung, S., Mak, S. and Lee, B., 2008. *Using a real-time integrated communication system to monitor the progress and quality of construction works*. Automation in Construction Vol 17, pp 749-757

Li, J. and Yang, H., 2017. *Research on Development of Construction Industrialization based on BIM Technology under the Background of Industry 4.0*. MATEC Web of Conferences 100, 02046

Liker, J., 2004. *The Toyota Way*. McGraw-Hill, Two Penn Plaza, New York

Lin, Y., Chang, J. and Su, Y., 2016. *Developing Construction Defect Management System Using BIM Technology in Quality Inspection*. Journal of Civil Engineering and Management Vol 22(7), pp 903-914

Machine Vision, 2018. *Machinevision.co.uk*. [online] Available at: <http://www.machinevision.co.uk/#sthash.mkD8iNPI.3dYuoIJQ.dpbs> [Accessed 20 January 2018]

Made Smarter, 2017. *Made Smarter Review*. [online] Available at: <https://assets.publishing.service.gov.uk/government/uploads/system/uploads/attachment_data/file/655570/20171027_MadeSmarter_FINAL_DIGITAL.pdf> [Accessed 24 October 2018]

Maguad, B., 2006. *The Modern Quality Movement: Origins, Development and Trends.* Total Quality Management and Business Excellence Vol 17 No. 2, pp 179-203

Martinez-Mesa, J., González-Chica, D., Duquia, R., Bonamigo, R. and Bastos, J., 2016. *Sampling: how to select participants in my research study.* An Bras Dermatol Vol 91(3), pp 326-330

McKinsey, 2016. *Imagining Construction's Digital Future.* [online] Available at: <https://www.mckinsey.com/industries/capital-projects-and-infrastructure/our-insights/imagining-constructions-digital-future> [Accessed 20 November 2017]

Mento, A., Jones, M. and Dirndorfer, W., 2002. *A change management process: Grounded in both theory and practice.* Journal of Change Management Vol 3(1), pp 45-59

Miles, M., Huberman, A. and Saldana, J., 2014. *Qualitative Data Analysis.* Sage Publications Ltd., Bonhill Street, London EC2

Mixamo, 2017. *Mixamo Animated 3D Characters.* [online] Available at: <https://www.mixamo.com/#/> [Accessed 9 November 2017]

Navon, R., 2005. *Automated project performance control of construction projects.* Automation in Construction Vol 14, pp 467-476

NBS, 2018. *BIM Levels explained.* [online] Available at: <https://www.thenbs.com/knowledge/bim-levels-explained> [Accessed 17 February 2018]

Oesterreich, T. and Teuteberg, F., 2016. *Understanding the implications of digitisation and automation in the context of Industry 4.0: A triangulation approach and elements of a research agenda for the construction industry.* Computers in Industry Vol 83, pp 121-139

Park, C., Lee, D., Kwon, O. and Wang, X., 2013. *A framework for proactive construction defect management using BIM, augmented reality and ontology-based data collection template.* Automation in Construction Vol 33, pp 61-71

PAS, 2013. *Publicly Available Specification 1192-2:2013.* [online] Available at: <https://shop.bsigroup.com/ProductDetail?pid=000000000030281435> [Accessed 26 November 2017]

Pickard, A., 2013. *Research Methods in Information.* Facet Publishing, Ridgmount Street, London

Pix 4D, 2017. *Pix 4D Drone Mapping and Photogrammetry Software.* [online] Available at: <https://pix4d.com/> [Accessed 9 November 2017]

Podfather, 2018. *Podfather PDA and back office systems.* [online] Available at: <https://www.thepodfather.com/> [Accessed 28 March 2018]

PwC, 2016. *Industry 4.0: Building the digital enterprise.* [online] Available at: <https://www.pwc.com/gx/en/industries/industries-4.0/landing-page/industry-4.0-building-your-digital-enterprise-april-2016.pdf> [Accessed 14 August 2018]

Rabionet, S., 2011. *How I Learned to Design and Conduct Semi-Structured Interviews: An Ongoing and Continuous Journey.* The Qualitative Report Vol 16:2, pp 563-566

Redwood, J., Thelning, S., Elmualim, A. and Pullen, S., 2017. *The proliferation of ICT and digital technology systems and their Influence on the dynamic capabilities of construction firms.* Procedia Engineering Vol 180, pp 804-811

Slack, N., Chambers, S., Johnston, R. and Betts, A., 2006. *Operations and Process Management.* Pearson Education Limited, Harlow, England

Smith, H. and Green, T., 1980. *Human Interaction with Computers.* Academic Press, London

Structure, 2018. *Structure Sensor.* [online] Available at: <https://structure.io/> [Accessed 18 January 2018]

University of Cambridge, 2017. *Getting virtual infrastructure models out of the computer and into the workspace.* [online] Available at: <http://www.eng.cam.ac.uk/news/getting-virtual-infrastructure-models-out-computer-and-workspace> [Accessed 15 November 2017]

Viewpoint, 2018. *Field View.* [online] Available at: <https://viewpoint.com/en-gb/products/viewpoint-field-view> [Accessed 15 October 2018]

Wadsworth, H., Stephens, K. and Godfrey, A., 2002. *Modern Methods for Quality Control and Improvement.* John Wiley & Sons, New York

Wang, J., Sun, W., Shou, W., Wang, X., Wu, C., Chong, H., Liu, Y. and Sun, C., 2015. *Integrating BIM and LiDAR for Real-Time Construction Quality Control.* Intelligent Robot Systems Vol 79, pp 417-432

Wang, L., 2008. *Enhancing construction quality inspection and management using RFID technology.* Automation in Construction Vol 17, pp 467-479

Wang, X. and Chong, H., 2015. *Setting new trends of integrated Building Information Modelling (BIM) for construction industry.* Construction Innovation Vol 15.1, pp 2-6

Whitton, N., 2007. *An investigation into the potential of collaborative computer game-based learning in Higher Education.* [online] Available at: <https://www.napier.ac.uk/~/media/worktribe/output-237315/whittonpdf.pdf> [Accessed 21 July 2018]

Woodhead, R., Stephenson, P. and Morrey, D., 2018. *Digital construction: From point solutions to IoT ecosystem.* Automation in Construction Vol 93, pp 35-46

APPENDICES

Appendix A

Interview Questions

Question 1 - What are the principal documents that bind the quality assurance processes together?

Question 2 - What are the changes required to existing processes in your preferred document management systems to enable fully digital quality applications?

Question 3 - Does the application of Building Information Modelling assist digitally connected quality assurance? Welcome to expand your response.

Question 4 - In the list of digital technologies indicated at the interview do you believe there are any missing applications appropriate for construction phase quality management?

Question 5 - How do we ensure staff and subcontractors reap the potential benefit of the digital technologies?

Question 6 - Your thoughts please for the future of digitally enabled quality assurance in the construction industry?

Question 7 - Which suite of these digital technologies indicated at the interview is the most appropriate for the next four years?

Appendix B
Digital Engineering Applications for Quality Assurance

Digital Engineering Applications for Quality Assurance				
Digital Application	Appropriate digital technologies for quality assurance? (indicated by Interview No.)	No. interviews process considered appropriate	Wide Use (applications appropriate by more than 75% participants)	Localised Implementation (appropriate by more than 50% participants)
Information technology enabler such as electronic document management system	01,02,03,04,05,06,07,08,09,10,11,12,13,14,15,16,17,18,19,20,21,22,23,24,25	25	✓	
Mobile phone and/or tablet as personal digital assistant	01,02,03,04,05,06,07,08,09,10,11,12,13,14,15,16,17,18,19,20,21,22,23,24,25	25	✓	
Building Information Modelling	01,02,03,04,05,06,07,08,09,10,11,12,13,14,15,16,17,18,19,20,22,23,24,25	24	✓	
Mobile and web application products	01,02,03,04,05,06,07,08,09,10,11,12,13,14,15,16,17,18,19,20,21,22,23,24,25	25	✓	
Digital inspection & test plan (ITP)	03,06,07,09,10,11,13,14,17,19,20,21,22,23,24,25	16		✓
Clash mitigation using BIM in construction phase	01,02,03,04,05,06,08,10,11,12,13,14,16,17,18,19,20,21,22,23,24,25	22	✓	
Real time performance information	01,02,03,04,05,06,07,08,09,10,11,12,13,14,15,16,17,18,19,20,21,22,23,24,25	25	✓	
Mixed reality when digital BIM modelling and real content co-exist	03,04,05,08,14,16,17,21,24	9		
Point clouds of as-built construction	01,02,03,04,05,06,07,08,09,10,11,12,13,14,15,17,18,19,20,21,23,24,25	23	✓	
Precise 3D vision to a mobile device	02,03,04,06,08,09,10,12,13,15,16,18,19,21,22,23,24,25	18	✓	
Sensors within progressive manufacturing	03,04,06,08,09,10,11,13,14,15,21,24	12		
Condition based monitoring	03,04,06,09,10,14,15,16,20,21,22,24,25	13		✓
3D geological models	03,04,06,08,10,11,13,14,16,19,21,23,24,25	14		✓
Character rigging and three-dimensional animation	03,04,06,09,13,14,17,20,23	9		
Business intelligence that reviews current media information	04,06,10,12,15,17,18,19,21,22	10		
Radio Frequency Identification	02,03,04,06,07,08,09,10,11,12,13,15,17,18,21	15		✓
Photogrammetric vision	02,03,04,05,06,07,08,09,10,14,15,16,17,20,23,25	16		✓
Machine vision	03,04,05,06,09,15	6		
Simultaneous localization and mapping	03,04,06,09,10,13,14,20,24	9		
2D (QR code) and 1-dimensional barcodes	01,02,03,04,05,06,07,08,09,10,11,12,13,14,15,16,17,18,19,20,21,23,24,25	24	✓	
Backpack mobile mapping	03,04,06,08,09,10,12,13,14,15,16,17,20,22,24,25	16		✓
Automated construction plant	02,03,06,14,21,24	6		

Publisher: Eliva Press SRL

Email: info@elivapress.com

Eliva Press is an independent publishing house established for the publication and dissemination of academic works all over the world. Company provides high quality and professional service for all of our authors.

Our Services:
Free of charge, open-minded, eco-friendly, innovational.

-All services are free of charge for you as our author (manuscript review, step-by-step book preparation, publication, distribution, and marketing).
-No financial risk. The author is not obliged to pay any hidden fees for publication.
-Editors. Dedicated editors will assist step by step through the projects.
-Money paid to the author for every book sold. Up to 50% royalties guaranteed.
-ISBN (International Standard Book Number). We assign a unique ISBN to every Eliva Press book.
-Digital archive storage. Books will be available online for a long time. We don't need to have a stock of our titles. No unsold copies. Eliva Press uses environment friendly print on demand technology that limits the needs of publishing business. We care about environment and share these principles with our customers.
-Cover design. Cover art is designed by a professional designer.
-Worldwide distribution. We continue expanding our distribution channels to make sure that all readers have access to our books.

www.elivapress.com